# PUZZLOONEY!

## RE4LLY R1D1CUL0US M4+H PU33LE5

### by Russell Ginns

Scientific American BOOKS FOR YOUNG READERS    W. H. FREEMAN AND COMPANY    NEW YORK

*I would like to thank the publisher,
Jacqueline Ball. How about that Al Nagy guy?
Didn't he do a swell job designing this book?
Sir Richard Weiss, Bill Basso, Dan Brawner,
Richard Lieser, Nathan Y. Jarvis:
Hey! Great job on the illustrations!
And of course, thanks to Lynn Brunelle.
She did oh-so-much work on this book,
and was very nice, too.*

*This book is dedicated to Lisa.*

**Copyright © 1994 Russell Ginns**

Printed in the U.S.A.

10 9 8 7 6 5 4 3 2 1

Library of Congress Cataloging-in-Publication Data
Ginns, Russell.
Puzzlooney: really ridiculous math puzzles/by Russell Ginns.
p. cm.
ISBN 0-7167-6532-2                    (Z)
1. Mathematical recreations—Juvenile literature.
[1. Mathematical recreations.] I. Title.         93-42178
QA95.G52 1994                                    CIP
793.7'4—dc20                                     AC

# WELCOME TO PUZZLOONEY,

A crazy collection of the world's wackiest math puzzles! Inside, you'll find bees, bugs, frogs, dogs, flytraps, trombones and pepperoni pizzas. Many of the puzzles are easy to solve, but you'll really have to use your noodle to answer others. This book will keep you puzzling and laughing until the cows come home. Now, why are you living with cows? That's your business.

As you solve your way through the book, you'll see several symbols that are there to guide you.

**Warning!**
**This puzzle is a tough one.**

**Solve this puzzle, and you'll get an answer to a corny riddle.**

**This puzzle's not too tough—if you can just find the trick to it.**

**The title of this puzzle has the same number of letters as there are in GEORGE WASHINGTON.**

Good luck, and remember: Do something about those cows, before it's too late!

# SUNDAE DRIVER

START

FINISH

MOO-LICIOUS

DING DING

When Albert Doubledip saw the Tyrannosaurus coming after him, he had to think fast! So he left a trail of ice cream cones to slow the monster down. Unfortunately, after Rex snacked on 20 scoops, he caught up with the truck. The big beast ate the entire supply of ice cream—plus the two rear tires.

**Can you find the path the dinosaur took?**

Start at the dinosaur and trace a path to the ice cream truck. Add up all of the ice cream scoops you pass along the way. You may not move diagonally, and cones do not count. The correct trail crosses over exactly 20 scoops.

After 25 years, the Grayola crayon company decided to add a new color to its collection. For the first time ever, their boxes will contain a color that isn't a shade of gray.

**Can you guess the name of their new hue?**

Read the three clues below. Then study all of the crayons. The one that fits all the clues is the newest color in Grayola's line.

# SOURCE OF A DIFFERENT COLOR

## CLUES

**1.** The name is more than nine letters long

**2.** There is a B in the name

**3.** The color is not "snowmobile"

# BEE-HIVE YOURSELF

It's easy to get lost inside a beehive.
The same goes for this puzzle. So don't
bumble, and wasp what you're doing.
Then, you'll buzz right through it.

**Only four spaces in this hive
contain numbers that don't touch a
space with the same number. Can you
spot them?**

# THE FINAL COUNTDOWN

The Neptune Explorer is ready to fire its rockets, but General Blastovski seems to have forgotten what to say. Help him find the numbers from ten to one so he can count down and get this mission started.

**Start at any ten and find a path that goes through all the numbers, in order, down to one.**

Be quick about it, though. Everyone's waiting!

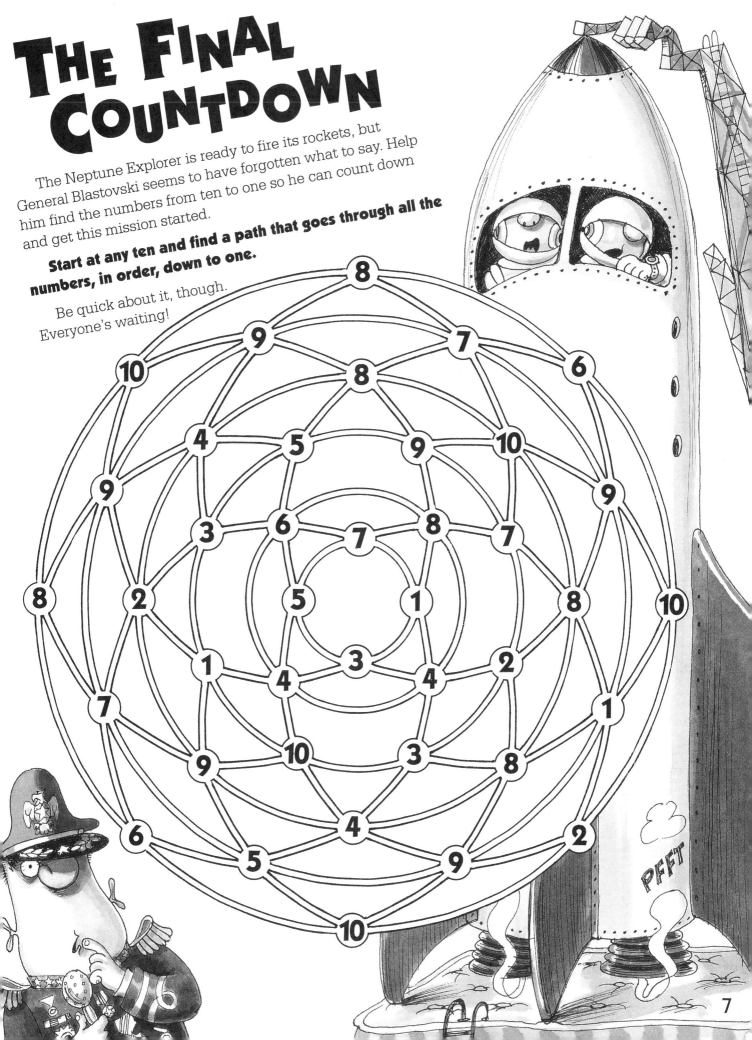

# THE LEAFY LADIES WHO LUNCH

25 flies flew into the experimental greenhouse.
- Esme ate three of them.
- Doris ate twice as many flies as one of the girls next to her did.
- Together, Agnes and Esme ate just as many flies as Doris did.
- Hilda ate three times as many flies as Agnes did.

**How many flies got out of there alive?**

DANGER
KEEP
OUT!

HILDA    DORIS    ESME    AGNES

# LET IT SLIDE

Lynn gave her trombone concert before a sold-out audience. But she didn't practice much, and she forgot to take the gum out of her mouth before she started playing. Solve this puzzle to find out what happened.

Every trombone is pointing directly at a musical note. Draw a straight line between all the pairs where the numbers by the trombone can be made to equal the number in the note by doing just one thing (adding, subtracting, multiplying, or dividing). The first one is done for you. The leftover letters will spell the answer to this riddle:

**What happened when Lynn played her trombone without practicing?**

11

On Monday, Hank Axlebend drove his truck to Nowheresville and went home again. The next day, he rode to Bubbly, then went to Snoosh and headed over to West Eastburg, where he spent the night. On Wednesday morning, he got up and drove down to Barnacle City. As he arrived, he noticed that he had traveled a total of 112 miles in three days.

**Hank always takes the shortest route possible between places. Can you figure out which city is his home?**

# KING OF THE ROAD

# SCOPE THIS OUT

Doctor Kleinbottle looked into his microscope and made an amazing discovery. Several hours later, he realized that he was looking at his tie. In the meantime, he came up with this puzzle:

**Look for all of the squares, triangles, and circles in this picture. Which shapes are the most numerous?**

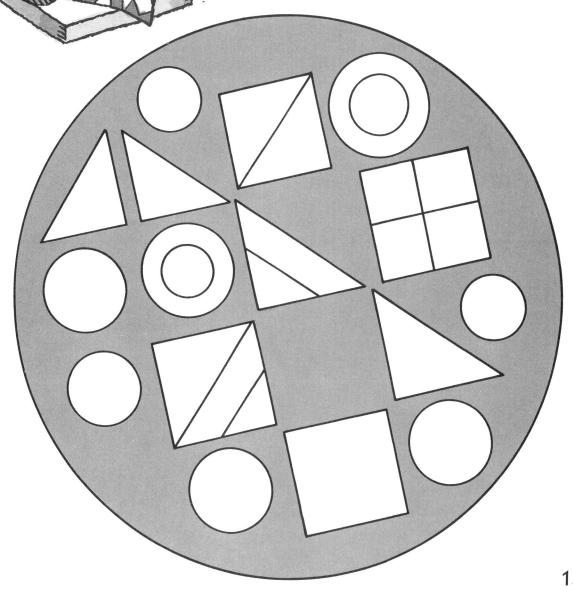

# LOOSE

*Norman's NEWT Ranch

| START | | | | |
|---|---|---|---|---|
| 1 | 2 | 3 | 1 | 2 |
| 2 | 5 | 2 | 3 | 6 |
| 3 | 7 | 1 | 8 | 5 |
| 4 | 2 | 3 | 7 | 8 |
| 2 | 6 | 4 | 6 | 2 |
| 3 | 3 | 5 | 3 | 1 |
| 4 | 6 | 3 | 4 | 5 |

Nine naughty newts escaped from Norman's newt ranch. So he grabbed his newt net and tried to round them up. This turned out to be a hard thing to do. You'll find that out, too, as you try to get all nine newts back together again.

+1

# NEWTS

| 3 | 4 | 5 | 6 | 7 | 8 |
|---|---|---|---|---|---|
| 8 | 7 | 6 | 4 | 5 | 4 |
| 6 | 5 | 4 | 5 | 7 | 3 |
| 3 | 5 | 6 | 7 | 3 | 4 |
| 3 | 4 | 3 | 5 | 2 | 5 |
| 6 | 5 | 4 | 6 | 4 | 8 |
| 6 | 7 | 8 | 7 | 8 | 9 |

Start with one newt in the top left corner and try to reach the space containing all nine newts. You can only enter a space that has a number that is one more or two less than the space you are on. You may not move diagonally. Good luck!

15

# FUNNY MONEY

Grab any one of these bills. Try to pick the one that's worth the most money. But wait! There are three important things to keep in mind:

- **A bill with a picture of a woman on it is worth triple the printed value.**
- **Divide the value by two if the printed value is an even number.**
- **A bill with a seven anywhere on it is worth nothing.**

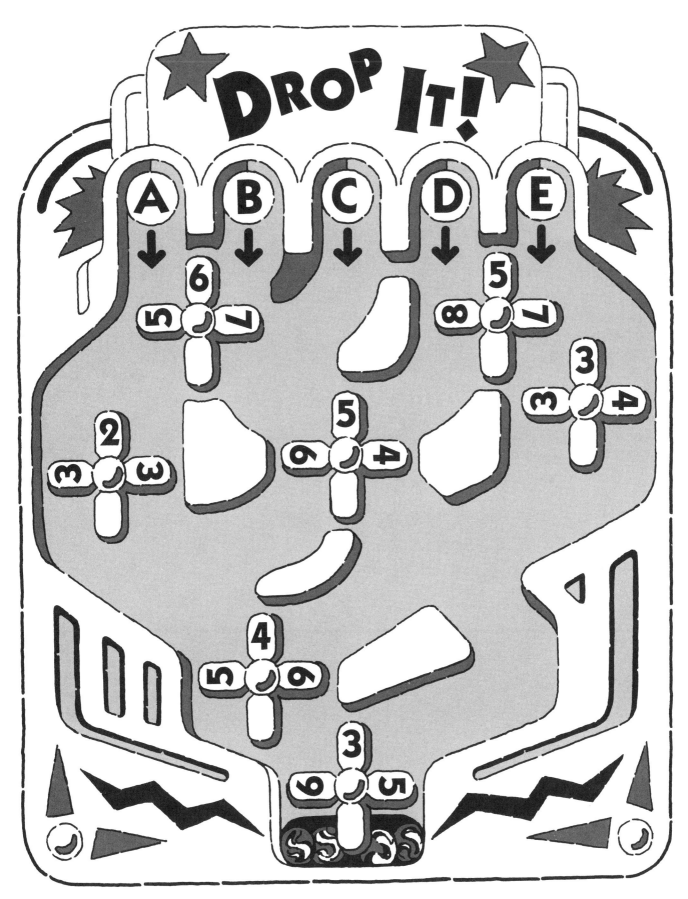

Depending on where you drop a marble into this machine, some spinners will move one notch and some will not. Can you guess which slot to drop it in so that the right-side-up numbers on **all** of the spinners in the machine will add up to 30?

# THE VEST OF TIMES

Lyle Houndstooth is one snappy dresser. He has only two fashion rules to follow:

- Never wear anything plaid with anything that has stripes.
- Never wear anything with pictures of the Eiffel Tower on it.

Below are all the shirts and pants in Lyle's closet.

**How many outfits can he create while sticking to his rules?**

# BLACK BOARD BUNGLE

**Can you use the symbols in the box at the bottom so that all four equations are correct?**

NOTE: You must use each one of them once and only once!

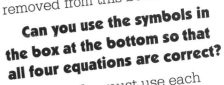

$$2 \quad 2 = 4$$

$$9 \quad 3 \quad = 2 \quad 1$$

$$1 \quad 7 = 8 + 4 \quad 5$$

$$1 + 2 \quad 3 \quad 9 - 3$$

$$\times \quad + \quad - \quad +$$

$$+ \quad = \quad \div$$

19

# TOO GROSS FOR

Answer: ___ ___   ___ ___ ___ ___ ___

20

# COMFORT

What happens if you wait too long to take out the trash? You get a smelly house and a few bugs. Big deal! But what happens if you NEVER take out the trash? You get this puzzle.

Start at the trash can that has only one fly. Then read all the cans in order, from the one with the least flies to the one with the most. They'll spell out the answer to this riddle:

**Why should you take your shopping cart to the dump?**

E

B

T

S

O

E

S

Y

E

After months of practice, Lynn was ready to play her trombone in public again. She remembered to take the chewing gum out of her mouth, too. This time, the concert was a big success. Solve this puzzle to find out why.

# FINE TUNED

Draw a line between trombones and notes if the three trombone numbers can be made to equal the number in the notes. (For example: 2+7-4=5.) The leftover letters will spell the answer to this riddle:

**Why was Lynn's concert so much better the second time?**

| S | Y | D | O | H | E |
| I | O | T | U | R | N |
| T | U | H | A | O | O |
| K | R | I | R | N | O |
| F | E | S | E | O | G |
| T | A | Y | W | E | S |

GUM →

22

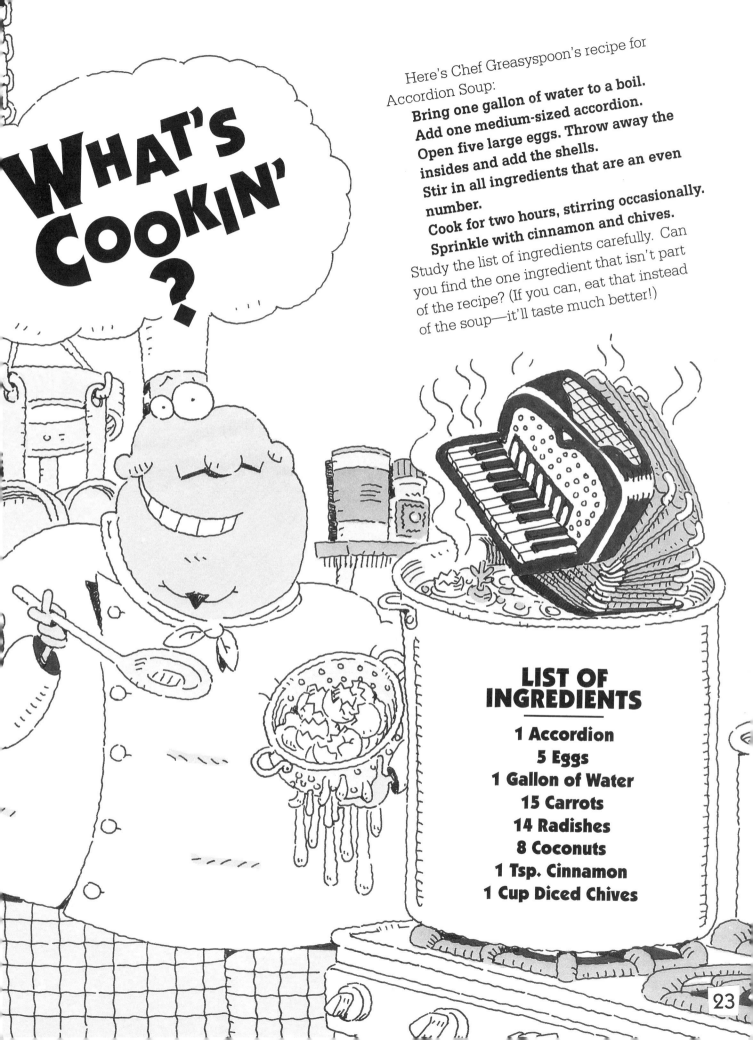

# WHAT'S COOKIN'?

Here's Chef Greasyspoon's recipe for Accordion Soup:

Bring one gallon of water to a boil.
Add one medium-sized accordion.
Open five large eggs. Throw away the insides and add the shells.
Stir in all ingredients that are an even number.
Cook for two hours, stirring occasionally.
Sprinkle with cinnamon and chives.

Study the list of ingredients carefully. Can you find the one ingredient that isn't part of the recipe? (If you can, eat that instead of the soup—it'll taste much better!)

## LIST OF INGREDIENTS

1 Accordion
5 Eggs
1 Gallon of Water
15 Carrots
14 Radishes
8 Coconuts
1 Tsp. Cinnamon
1 Cup Diced Chives

# WASHINGTON OR BUST

1766–1809  1729–1797  1691–1798  1730–1788

1741–1805  1730–1798  1732–1799  1733–1797

George Washington couldn't tell a lie. But "Honest" Abe, the statue salesman sure can! He claims that all eight of these Washington statues are authentic and accurate. You can tell that's not true, because you know three things:

- George Washington was born before 1733.
- He died after 1797.
- He didn't wear glasses.

**Can you find the one authentic bust of the father of our country?**

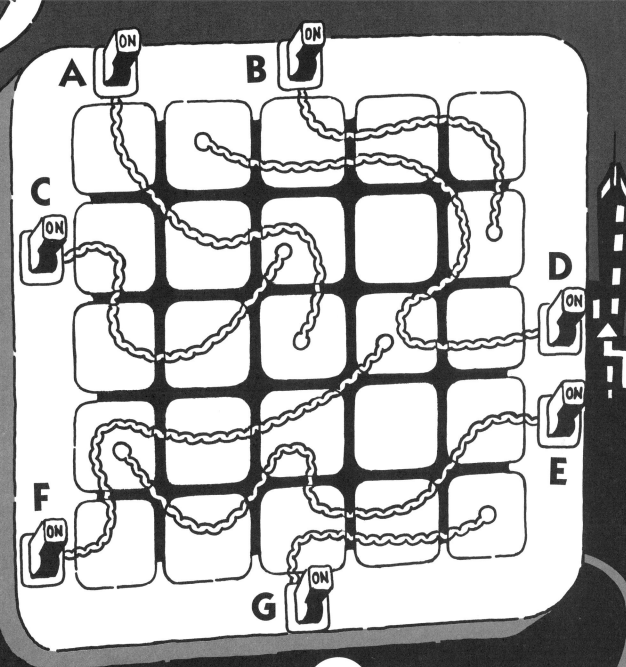

# LIGHTS OUT

The cost of electricity is going way up. So, to save money, the folks at Squares, Inc. are changing their glowing window sign from a big square to an X. Help them turn off some of their lighted panels and help them stay in business. When a switch is thrown, every square that its wire runs through will shut off.

**Which three switches need to be triggered so the leftover lit-up squares will form an X?**

# TEST

Julia likes to stand in front of the appliance store window so she can watch all her favorite TV talk shows at the same time. Of course, if she didn't watch so much television, maybe she'd notice that there's a maze here.

Start with the big TV at the top and try to find a path all the way to the bottom. Move from screen to screen, one TV at a time. You must always move in the direction that the person on the screen is looking. If the person is looking straight ahead, then you may go in any direction you like.

(You may not move diagonally.)

# TWIST AND SHOUT

It's time for the pre-semi-half-quarter-finals in the Twisted Valley gymnastics tournament.

**Can you figure out which athlete will get the highest score for this round?**

Read each gymnast's routine carefully. Then read the judges' guidelines. The gymnast with the highest total score is the winner.

## JUDGES' RULES:

- Score three points for each sneeze-flip, winch-twist, or neck-knock.
- Score ten points for each squeaking turnabout.
- Divide the final score by two if you fall down at any point.

> I will do two sneeze-flips, two neck-knocks, fall down, and do one squeaking turnabout.

> I will do a winch-twist, a neck-knock, and fall down.

> I will now do two sneeze-flips and one winch-twist. Then I will fall down. After that, I will perform another perfect winch-twist.

> I will do three winch-twists.

> I will do a squeaking turn-about. Then I will lose my balance and fall down.

# THEIR BARK IS OKAY, BUT...

No one likes to deliver the mail in Pitbulton, Ohio. There are just too many vicious dogs on long leashes. In fact, there's only one mailbox in the town that a mail carrier can get to without facing down a ferocious Fido. Can you spot the one safe box? Study the map carefully and try to estimate how far each leash can go.

**Pick the mailbox that is safely beyond the reach of all four dogs.**

# GOTCHA!

Jeremy's prize bullfrog won the local jumping contest. Then he jumped out the door and got away! Can you help catch him in time for the state championship?

Start at the sign and hop one space to the right. Each time you land on a new square, follow the numbered arrow and jump that many spaces in a straight line. Keep going until you catch that pesky frog.

# COOKIES AND CREEPS

Hungry? Then grab some of these cookies. In fact, try to get as many chocolate chips as you can. There is only one rule: All the cookies you take must have the same number of chips on them.

**WARNING: All the cookies that have an odd number of chips on them are actually rare fakemtrixus beetles. If you take any of them, you lose!**

**Which cookies should you choose to get the most chocolate chips?**

31

# COLLECT CALL

Clyde was busy writing out his latest novel, when the phone rang. He picked up the receiver and got an odd message:

## 4-3-8 2 8-9-7-3-9-7-4-8-3-7!

**Using the buttons on his telephone as a guide, he was able to figure out the meaning of the strange call. Can you?**

Sylvia's skating out of control in the ice rink and she can't stop! Can you figure out a way that she can skate over to the sofa without sliding out of sight? Start at the top, where Sylvia is headed straight down. Each time you reach a pole, you may change directions (up, down, left, or right). Otherwise, you have to keep moving in a straight line. Keep going until you reach the sofa safely. If you reach the edge of the rink, you lose!

# ON THIN ICE

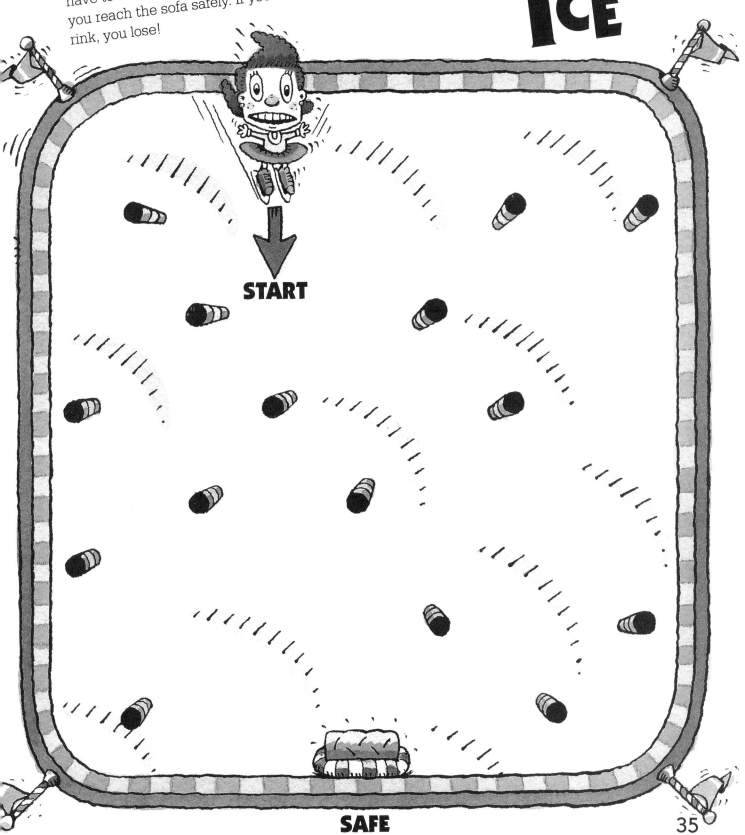

START

SAFE

# ANCHOVY EXPRESS

It's Howie's first day on the job. But being a delivery boy for Pizza Cave isn't going to be so easy. In fact, he might not even make his first delivery.

**Can you find the only route that Howie can take to drop off his pizza?**

Hurry! It's getting cold.

# UP AND AWAY!

- Janet and Martin took off in their hot air balloon from Feldronia.
- They sailed north until they reached a country approximately twice the size of the one they left.
- Then they floated to the smallest nation on that country's border.
- From that country, they drifted to the only other place on the map with the same number of letters in its name.
- They sailed east until they reached an island and landed safely.

**Where did they land?**

UPPER THOOM

LOWER THOOM

REPUBLIC of BUNJI

FLOSSO

FLAIRPEN

TIBIA

BOTT

FLEAZE

ULNA

FELDRONIA

BRICKLEY

N

E

W

S

GERMA

Greg Simpson woke to discover that he was slowly changing into a gigantic beetle! Fortunately, his wife had a loaded camera nearby. She captured the whole amazing event on film.

**Study the pictures below. Can you put them in order?**

# Beauty and the Beetle

FLOOR 1
ATFISH & HARKS

FLOOR 2
RAY, EEL, & MACKEREL

FLOOR 3
PIRANHA & BARRACUDA

FLOOR 4
TROUT & LUNGFISH

FLOOR 5
SALMON, SWORDFISH, & BASS

# GOING UP?

Sheldon and Marian went to the World Fish Center. But as soon as they climbed into the elevator, they forgot what kind of fish they wanted to buy. Solve this puzzle and help them bring home the fish of their dreams.

Read the three clues below and study the picture.

**Can you figure out which fish is the one they want?**

## CLUES:

1. Sheldon and Marian's fish does not have an S in its name.

2. Their fish is on an odd-numbered floor.

3. They really don't want a piranha.

39

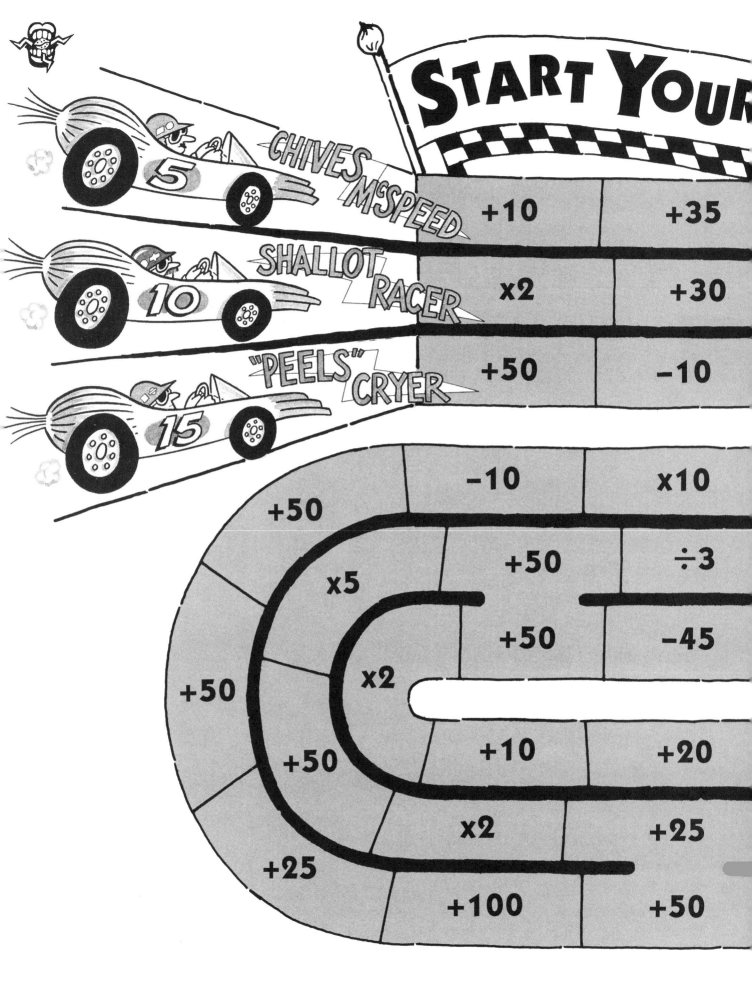

# ONIONS

Welcome to the Onionopolis 500, the world's first onion-powered auto race. Can you lead one of these scallion-speedsters to victory?

Start with any of the drivers and use the number on his or her car. Then trace a path toward the finish line. You may switch lanes whenever there is an opening, but you may never go backward. As you move along, add, subtract, multiply, or divide according to the space you enter. You win the race if you can cross the finish line with a value of exactly 500. Ready, set, go!

| x2 | | ÷5 |
| x2 | x2 | |
| +25 | +25 | x6 |
| | | +5 |

| +25 | +25 | x8 |
| +100 | +50 | ÷2 |
| ÷5 | −10 | −5 |

| +30 | | +40 |
| −100 | −100 | −100 |
| x5 | x2 | +50 |

500

# TOPPING THE CHARTS

At midnight on Wednesday, Howie at the Pizza Cave got orders for six pepperoni pizzas. All of the people below were listening to WIMP radio at that time.

- When Rob hears "Funky Town," he calls Zelda on the phone and plays "Greensleeves" for her. Then he orders two pepperoni pizzas.
- Whenever Zelda hears "Pop Goes the Weasel," she orders two pepperoni pizzas and bites Scott on the toe.
- Each time Scott gets bit on the toe or hears the song "Greensleeves," he orders a pepperoni pizza.
- When Davida hears the song "Greensleeves," she calls Jill and Scott on the phone and plays them "Pop Goes the Weasel."
- Every time Jill hears "Pop Goes the Weasel," she calls up Rob and plays "Funky Town," then orders a pepperoni pizza.

**Can you figure out which song must have been playing just then?**

# EYES ON THE PIES

The citizens of Crustburg, New Jersey, held their annual pizza baking contest and these pizzas came in last place. How can you tell? The judges left most of these five pies untouched.

All of the pies had the same area when the contest began.

**Can you figure out which is absolutely the worst pizza of them all?**

Chicago-Style Deep Dishwater Pan Pizza

Anchovies, Artichokes, Avocado, & Aspirin

Candied Carrots & Clams

Sandpaper with Extra Cheese

Aunt Edna's Oatmeal & Asparagus

# BOWLING FOR DISASTER

Carmen threw her bowling ball the wrong way and knocked down six of her friends. She thought about it for a while and decided that it was much more fun than plain old bowling for pins. LOOK OUT!

**Can you find the two bowling balls that will knock down the most people without touching any bowling pins?** (You can use a ruler.)

# ROUTE, ROUTE, ROUTE FOR THE HOME TEAM

Below is a map of Strikeout County and this week's baseball schedule. Teams always take the shortest route to an away game and come right home afterward.

**Can you figure out which team will have traveled the most miles by the end of the week?**

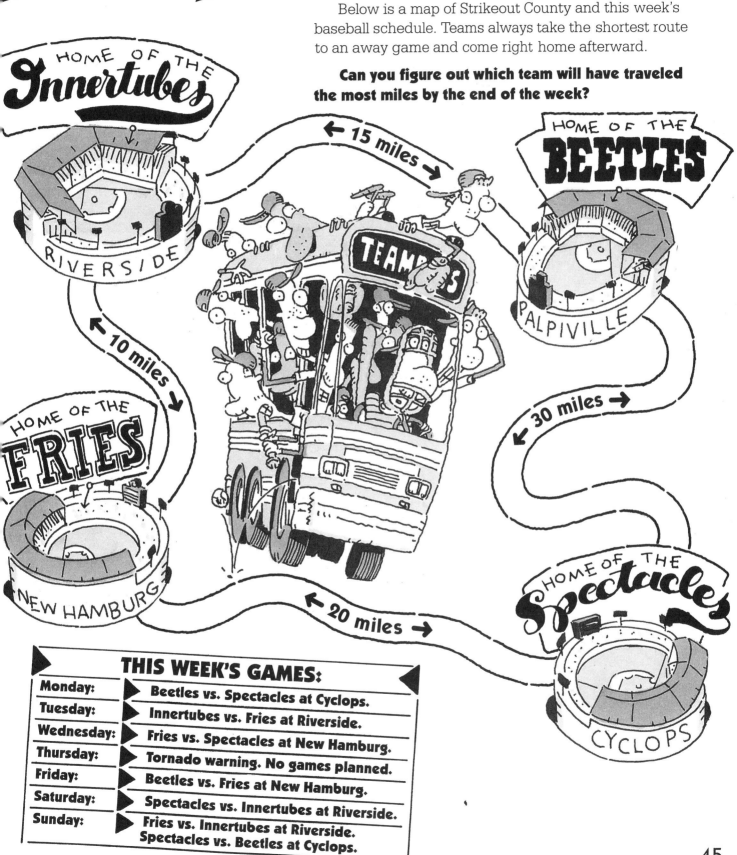

HOME OF THE *Innertubes*

RIVERSIDE

← 15 miles →

HOME OF THE BEETLES

PALPIVILLE

← 10 miles →

HOME OF THE FRIES

NEW HAMBURG

TEAMBUS

← 30 miles →

← 20 miles →

HOME OF THE *Spectacles*

CYCLOPS

## THIS WEEK'S GAMES:

| | |
|---|---|
| **Monday:** | Beetles vs. Spectacles at Cyclops. |
| **Tuesday:** | Innertubes vs. Fries at Riverside. |
| **Wednesday:** | Fries vs. Spectacles at New Hamburg. |
| **Thursday:** | Tornado warning. No games planned. |
| **Friday:** | Beetles vs. Fries at New Hamburg. |
| **Saturday:** | Spectacles vs. Innertubes at Riverside. |
| **Sunday:** | Fries vs. Innertubes at Riverside. Spectacles vs. Beetles at Cyclops. |

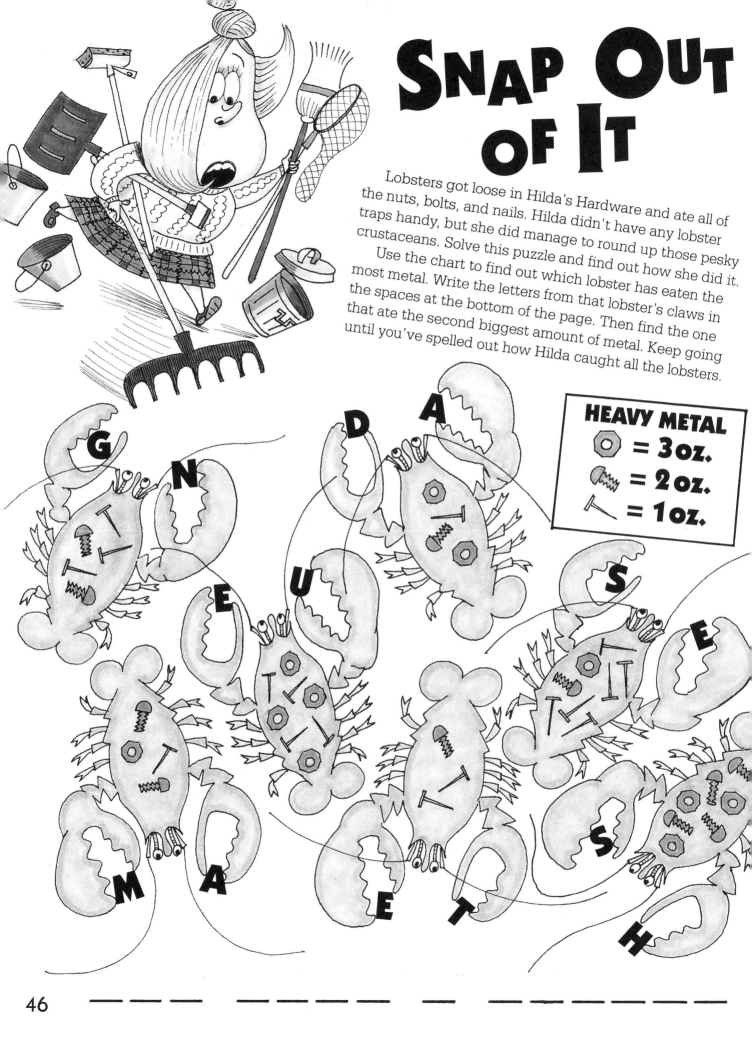

# SNAP OUT OF IT

Lobsters got loose in Hilda's Hardware and ate all of the nuts, bolts, and nails. Hilda didn't have any lobster traps handy, but she did manage to round up those pesky crustaceans. Solve this puzzle and find out how she did it.

Use the chart to find out which lobster has eaten the most metal. Write the letters from that lobster's claws in the spaces at the bottom of the page. Then find the one that ate the second biggest amount of metal. Keep going until you've spelled out how Hilda caught all the lobsters.

**HEAVY METAL**
= 3 oz.
= 2 oz.
= 1 oz.

# OFF THE SHELF

Al Snackem pulled three items from his kitchen cupboard and noticed something unusual. All of the food remaining on each shelf added up to exactly the same weight.

**Can you guess which three products he took away?**

Gettysburg Cheese
AT LEAST 87 YEARS OLD
WEIGHT: 14 oz.

NOT VERY GOOD Cola
WEIGHT: 10 oz.

DIET NOT VERY GOOD Cola
WEIGHT: 8 oz.

Wigglies
SNAKE SHAPED JELLY TREATS
WEIGHT: 10 oz.

ATOMIC VOLCANO LAVA TREATS
WEIGHT: 9 oz.

LEND ME YOUR EARS CORN CLUSTERS
WEIGHT: 16 oz.

GREY GOOP FANCY MUSTARD
WEIGHT: 3 oz.

WEIGHT: 2 oz.

NEW SMALLER BITS
ITTY BITTY COOKIE BITS

Nothing O's
WEIGHT: 1 oz.
AIR FLAVORED

WEIGHT: 20 oz.

Bell Bar
chocolate coated

KRAZY FRUITZ
APPLE TOMATO BANANA FLAVORED
WEIGHT: 9 oz.

COUCHEES
SOFA FLAVORED POTATO SNACKS
WEIGHT: 19 oz.

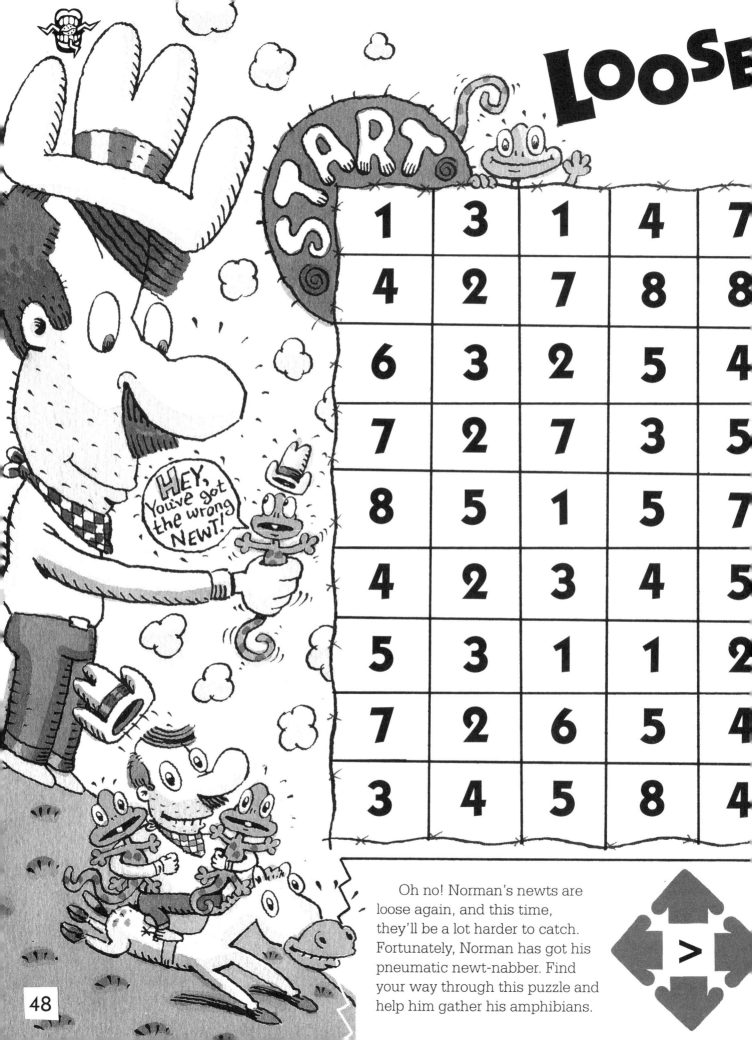

LOOSE

START

| 1 | 3 | 1 | 4 | 7 |
| 4 | 2 | 7 | 8 | 8 |
| 6 | 3 | 2 | 5 | 4 |
| 7 | 2 | 7 | 3 | 5 |
| 8 | 5 | 1 | 5 | 7 |
| 4 | 2 | 3 | 4 | 5 |
| 5 | 3 | 1 | 1 | 2 |
| 7 | 2 | 6 | 5 | 4 |
| 3 | 4 | 5 | 8 | 4 |

HEY, You've got the wrong NEWT!

Oh no! Norman's newts are loose again, and this time, they'll be a lot harder to catch. Fortunately, Norman has got his pneumatic newt-nabber. Find your way through this puzzle and help him gather his amphibians.

# NEWTS PART II

| | | | | |
|---|---|---|---|---|
| 3 | 2 | 8 | 4 | 2 |
| 5 | 6 | 7 | 2 | 1 |
| 3 | 2 | 4 | 3 | 1 |
| 6 | 3 | 1 | 7 | 3 |
| 2 | 4 | 5 | 1 | 5 |
| 3 | 4 | 1 | 3 | 6 |
| 2 | 5 | 6 | 2 | 7 |
| 3 | 6 | 2 | 5 | 8 |
| 2 | 5 | 7 | 8 | 9 |

OR ½

This time, you can only move to a new square if it has a greater number of newts than the square you are on OR if it has exactly half of your current newts. Again, you may not move diagonally. Happy hunting!

49

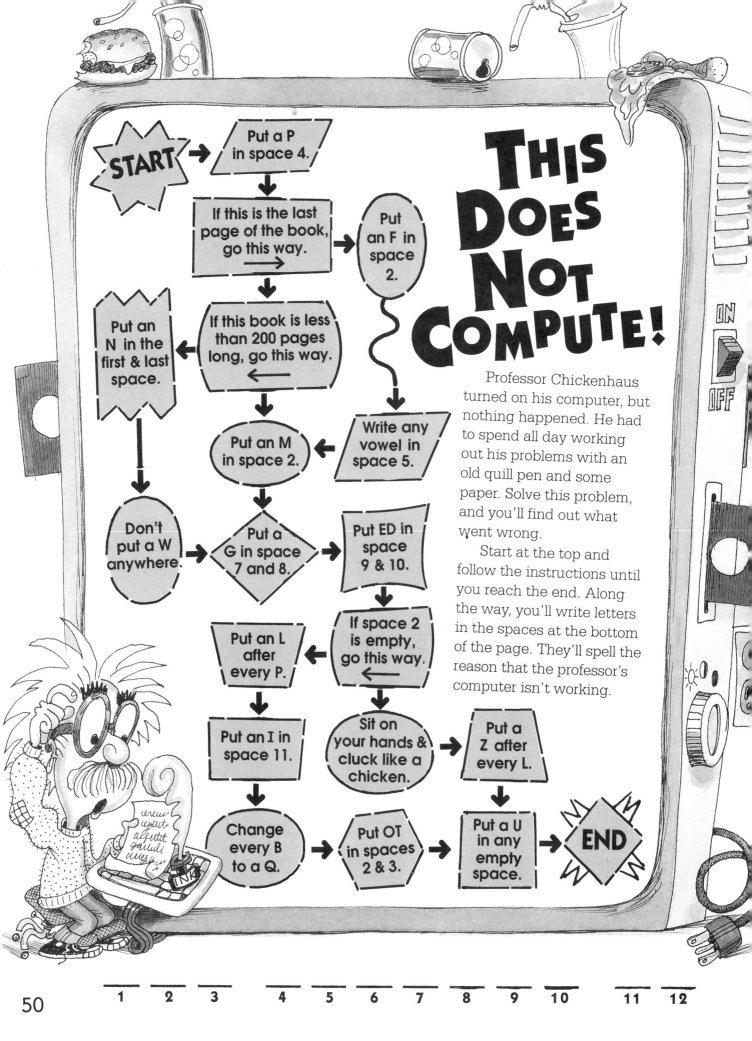

# THIS DOES NOT COMPUTE!

START → Put a P in space 4.

If this is the last page of the book, go this way. →

Put an F in space 2.

If this book is less than 200 pages long, go this way. ←

Put an N in the first & last space.

Write any vowel in space 5.

Put an M in space 2.

Don't put a W anywhere.

Put a G in space 7 and 8.

Put ED in space 9 & 10.

If space 2 is empty, go this way. ←

Put an L after every P.

Put an I in space 11.

Sit on your hands & cluck like a chicken.

Put a Z after every L.

Change every B to a Q.

Put OT in spaces 2 & 3.

Put a U in any empty space.

END

ON
OFF

Professor Chickenhaus turned on his computer, but nothing happened. He had to spend all day working out his problems with an old quill pen and some paper. Solve this problem, and you'll find out what went wrong.

Start at the top and follow the instructions until you reach the end. Along the way, you'll write letters in the spaces at the bottom of the page. They'll spell the reason that the professor's computer isn't working.

___  ___  ___  ___  ___  ___  ___  ___  ___  ___  ___  ___
1    2    3    4    5    6    7    8    9   10   11   12

# SOMETHING FISHY

What do you get when you mix a school of spotted snicklefish with a swarm of striped sand-swishers? You get this puzzle!

**Study these fish carefully. Can you flip just two fish and make it so the east- and west-heading stripes are equal to each other, and the east- and west-heading spots are equal to each other too?**

(The spots and stripes aren't the same number.)

# CLOSE, BUT NO CHOCOLATE

Lisa opened her candy bar and found a lucky silver ticket. She won a free trip to the chocolate factory! Unfortunately, she got on the wrong bus and wound up at the cigar factory instead.

**Can you help her find her way out?**

Start at the center and find a path to the exit. You can move into any space that touches the one you are on, just as long as it contains an equal number of boxes and cigars.

HAVASTOGIE CIGAR CO.

START

LOADING

EXIT

PIG, PIGGER, PIGGEST

At this year's pig contest, the pigs were so big that they broke the scales. Fortunately, the judges quickly rigged up a different way to weigh them.

**Study the pictures below and see if you can decide which is the biggest of the pigs.**

53

# PETALS TO THE METAL

Mr. Snoozem fell asleep at the wheel of his power mower and rolled right into his petunia patch.

**Can you find the path out of the garden that destroys the fewest flowers?**

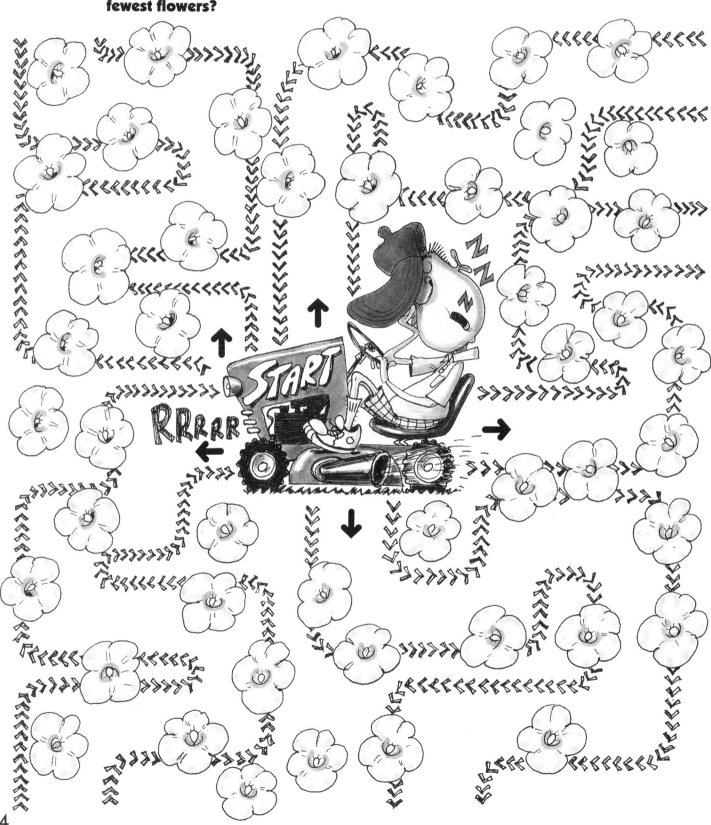

# RAZE THE ROOF

The space aliens who crash-landed into Martin's living room kindly offered to repair his ceiling. Unfortunately, the place is such a mess that they're not quite sure which pieces to use.

**Can you find the correct seven pieces needed to make Martin's roof whole again?**

# How Sweet It Is!

Lisa finally made it to the chocolate factory, and she wasted no time getting down to business. She ate thirty-six bags of chocolate chips before she passed out from chocolatosis. While we wait for her to wake up, solve this puzzle.

Draw lines to connect all the pairs of chocolate chips listed in the box. Then read the letters in the path you've just made. They'll spell out the answer to this riddle:

**What should you use to write down cookie recipes?**

| | | | |
|---|---|---|---|
| 2—8 | 3—9 | 9—10 | 8—14 |
| 10—16 | 14—15 | 15—21 | 16—22 |
| 20—21 | 20—26 | 22—28 | 27—28 |
| 28—29 | 29—30 | 26—32 | 32—33 |
| 33—34 | 34—35 | 35—36 | |

# TRIPLE TROUBLE

So! You think you've solved all of the puzzles in this book? Well, you're not finished until you solve this three-level challenge. Are you up to it? Here's what to do:

CLAP    CLAP

## STEP 1: WHAT'S MY LINE?

Read all of the numbers in each strip and try to find out each strip's pattern. Write the missing numbers in the empty spaces.

SQUEAK SQUISH

BE SURE AND TRY OUR CREAM OF CONCERTINA!

OULDN'T BE ROUDER

TOP SOUP

ACCORDIAN AND NOODLE

HARMONICA AND SPLIT PEA

STSOCKS

1 2 1 3 1 4 __ __ __ __

2 4 6 8 10 __ __ __

1 3 2 4 3 __ __ __

1 2 4 7 11 16 __ __ __

1 1 3 3 5 5 __ __ __ __

55 22 55 22 __ __ __ __

16 61 27 72 38 83 __ __

1 2 1 2 3 1 2 3 4 __ __ __

5 10 15 20 __ __ __

1 2 3 5 7 11 __ __ __

TURN THE PAGE FOR STEP 2 ➡

# TRIPLE TROUBLE

## STEP 2: NUMBER SEARCH

Hunt for all the numbers you've just written and cross them out. For example, if you've just filled in 12, 14, 16 and 18, they will appear somewhere in order in this grid. The series of numbers may go up, down, across, backward, or diagonally.

| 13 | 17 | 19 | 23 | 99 | 98 | 11 | 80 |
|----|----|----|----|----|----|----|----|
| 1  | 5  | 1  | 6  | 5  | 60 | 11 | 5  |
| 55 | 22 | 55 | 4  | 31 | 51 | 9  | 4  |
| 52 | 63 | 3  | 21 | 64 | 32 | 9  | 6  |
| 71 | 2  | 33 | 94 | 49 | 8  | 7  | 5  |
| 1  | 12 | 14 | 16 | 18 | 15 | 7  | 7  |
| 10 | 62 | 25 | 30 | 35 | 40 | 45 | 50 |
| 61 | 22 | 29 | 37 | 46 | 95 | 96 | 97 |

# TRIPLE TROUBLE

DINO-MITE!

CERTIFICATE OF MERIT

YOU EARNED IT!

## STEP 3: CROSS EM' OUT

Circle all of the leftover numbers in the grid. Then look for them in the boxes below and cross out all of the ones you find. When you've finished, the leftover letters will have a special message for you—and you will have earned it!

THAT'S FOR SURE.

| 60 S | 31 K | 12 Y | 13 O | 51 N |
|------|------|------|------|------|

| 21 F | 20 U | 64 L | 34 A | 48 R | 32 D |
|------|------|------|------|------|------|

| 39 E | 44 A | 10 T | 11 P | 63 O | 38 U | 62 B |
|------|------|------|------|------|------|------|

| 61 A | 23 Z | 37 Z | 95 Y | 36 L | 99 E | 4 O | 98 N |
|------|------|------|------|------|------|------|------|

| 80 U | 26 O | 18 N | 52 A | 33 M | 35 E |
|------|------|------|------|------|------|

| 28 Y | 96 E | 8 A | 27 C | 19 H | 15 C |
|------|------|------|------|------|------|

| 7 A | 30 M | 71 O | 3 P | 97 A |
|-----|------|------|-----|------|

I'M DONE!

FINISH

WINNER

MOO LICIOUS

TO THE WINNER GOES THE SCOOPS!

59

# ANSWERS FOR PUZZLOONEY

## P.4 **SUNDAE DRIVER**

## P.5 **SOURCE OF A DIFFERENT COLOR**
**Answer:** Burnt Lumber

## P.6 **BEE-HIVE YOURSELF**

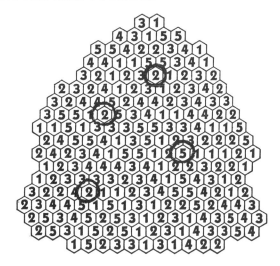

## P.7 **THE FINAL COUNTDOWN**

## P.8 **I WANT MY NTV**

| | | |
|---|---|---|
| A = 6 | D = 9 | G = 4 |
| B = 3 | E = 2 | H = 5 |
| C = 1 | F = 7 | I = 8 |

## P.10 **THE LEAFY LADIES WHO LUNCH**
4 Flies

## P.11 **LET IT SLIDE**

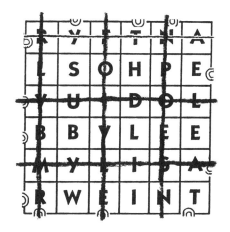

**Answer:** SHE BLEW IT

## P.12 **KING OF THE ROAD**
West Eastburg

## P.13 **SCOPE THIS OUT**
There are more circles than any other shape. (Did you forget to count the outline of the picture?)

## P.14 **LOOSE NEWTS**

P.16 **FUNNY MONEY**
Frieda Bankle 3rd

P.17 **DROP IT!**

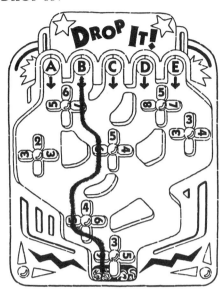

P.18 **THE VEST OF TIMES**
Lyle has 18 different outfits to choose from.

P.19 **BLACKBOARD BUNGLE**

2 + 2 = 4                     1 x 7 = 8 + 4 - 5

9 ÷ 3 = 2 + 1               1 + 2 + 3 = 9 - 3

P.20 **TOO GROSS FOR COMFORT**
**Answer:** TO BUY SOME GROSS-ERIES

P.22 **FINE TUNED**

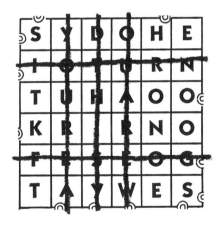

**Answer:** SHE TOOK NOTES

P.23 **WHAT'S COOKIN'?**
**Answer:** Carrots

P.24 **WASHINGTON OR BUST**

P.25 **LIGHTS OUT**
Switches C, D, and E.

P.26 **SCREEN TEST**

P.28 **TWIST AND SHOUT**
**Answer:** #1 will win with 11 points.

P.29 **THEIR BARK IS OKAY, BUT...**

61

## P.30 GOTCHA!

## P.31 COOKIES AND CREEPS
Grab the two cookies that have eight chips each.

## P.32 PARTY ON!

Numbers won't solve half our problems. Numbers will solve all our problems!

I plan to create new numbers where, today, there are only doodles and question marks.

The number party headquarters is in a big fancy building with the best staff I've ever seen.

Don't worry, everybody. The Number Party is here and you can count on us.

Vote for me, my good friends, and I promise there will be a digit on every doorknob and a number in every shoe!

Our commercials will be seen in every movie theater and living room across the country. Then, everybody will hear about our party.

We need numbers on all our walls, floors, and doors, even if that means we have to write on them with chalk.

## P.34 COLLECT CALL
**Answer:** GET A TYPEWRITER!

## P.35 ON THIN ICE

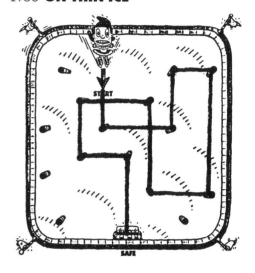

## P.36 ANCHOVY EXPRESS

## P.37 UP AND AWAY!
Janet and Martin landed in ULNA.

## P.38 BEAUTY AND THE BEETLE

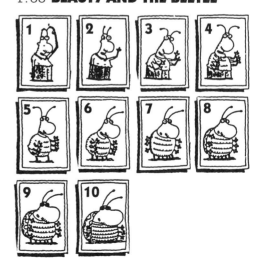

## P.39 GOING UP?

They want to buy a barracuda.

## P.40 START YOUR ONIONS

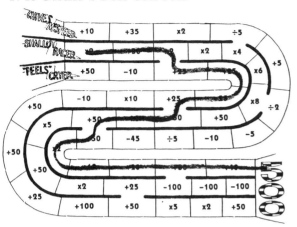

## P.42 TOPPING THE CHARTS

Answer: WIMP was playing "Pop Goes the Weasel."

## P.43 EYES ON THE PIES

The worst pizza (the most left) is the Chicago-Style Deep Dishwater Pan Pizza.

## P.44 BOWLING FOR DISASTER

## P.45 ROUTE, ROUTE, ROUTE FOR THE HOME TEAM

**Answer:** The Beetles

## P.46 SNAP OUT OF IT

**Answer:** SHE USED A MAGNET

## P.47 OFF THE SHELF

Al took the Gettysburg Brand Cheese, Grey Coop Brand Fancy Mustard, and the Bell Bar.

## P.48 LOOSE NEWTS, PART II

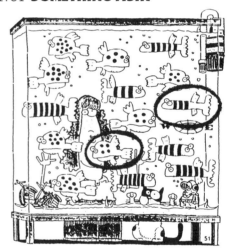

## P.50 THIS DOES NOT COMPUTE!

**Answer:** NOT PLUGGED IN

## P.51 SOMETHING FISHY

## P.52 CLOSE, BUT NO CHOCOLATE

63

## P.53 **PIG, PIGGER, PIGGEST**

Anthony is the biggest pig.

## P.54 **PETALS TO THE METAL**

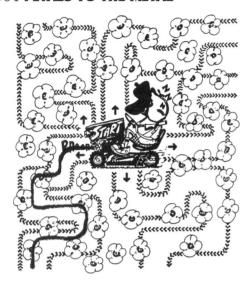

## P.55 **RAZE THE ROOF**

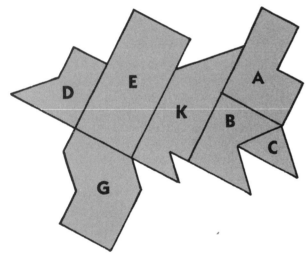

## P.56 **HOW SWEET IT IS!**

**Answer:** CHALK-OLATE

## P.57 **TRIPLE TROUBLE**

1 2 1 3 1 4 *1 5 1 6*
2 4 6 8 10 *12 14 16 18*
1 3 2 4 3 *5 4 6 5 7*
1 2 4 7 11 16 *22 29 37 46*
1 1 3 3 5 5 *7 7 9 9 11 11*
55 22 55 22 *55 22 55*
16 61 27 72 38 83 *49 94*
1 2 1 2 3 1 2 3 4 *1 2 3 4 5*
5 10 15 20 *25 30 35 40 45 50*
1 2 3 5 7 11 *13 17 19 23*

**Answer:** YOU ARE A PUZZLOONEY CHAMP!

64